GHOST PLANT

ASHLEY GISH

look!

GHOST PLANT

CREATIVE EDUCATION • CREATIVE PAPERBACKS

Published by Creative Education
and Creative Paperbacks
P.O. Box 227, Mankato, Minnesota 56002
Creative Education and Creative Paperbacks are imprints of
The Creative Company
www.thecreativecompany.us

Design and art direction by Blue Design
Edited by Kremena Spengler

Photographs by Dreamstime/Brian Lasenby, 17; Getty Images/Christopher Laszlo Bonis / 500px, 22, ilbusca, 23, James L. Amos, 9, Scott T. Smith, 2; Shutterstock/Storyjin, cover, 1, 23; Wikimedia Commons/Ayotte, Gilles, 20, Benny Mazur, 10, brewbooks, 15, Doug McGrady, 21, Douglas Goldman, 11, 13, 14, Internet Archive Book Images, 5, Jon Sullivan, 16, Jrg1000, 18–19, Provincial Archives of Alberta, 12, USGS Bee Inventory and Monitoring Lab, 8, Virginia State Parks, 6–7, Will Brown, 14–15
Every effort has been made to contact copyright holders for material reproduced in this book. Any omissions will be rectified in subsequent printings if notice is given to the publisher.

Library of Congress Cataloging-in-Publication Data
Names: Gish, Ashley, author.
Title: Ghost plant / Ashley Gish.
Description: Mankato, Minnesota : Creative Education and Creative Paperbacks, [2026] | Series: Look! | Includes bibliographical references and index. | Audience: Ages 6–9 | Audience: Grades 2–3 | Summary: "A see-through plant that needs no sunlight and feeds on a fungus? Elementary-level readers will learn all about the ghost plant with this striking introduction to the unusual plant's life cycle"– Provided by publisher.
Identifiers: LCCN 2024044293 (print) | LCCN 2024044294 (ebook) | ISBN 9798889895886 (library binding) | ISBN 9781682777541 (paperback) | ISBN 9798889896685 (ebook)
Subjects: LCSH: Indian pipe (Plant)–Juvenile literature.
Classification: LCC QK495.E68 G53 2026 (print) | LCC QK495.E68 (ebook) | DDC 583/.93–dc23/eng/20250113
LC record available at https://lccn.loc.gov/2024044293
LC ebook record available at https://lccn.loc.gov/2024044294

Printed in India

TABLE OF CONTENTS

The ghost plant is nearly see-through—like a ghost! Its drooping flowers grow on scaly stalks. It lives in North and South America and Asia.

Ghost plants are not like most other plants. They do not get energy from sunlight. They grow by feeding on a **fungus**.

fungus: a living thing that gets energy from dead and rotting plants

The ghost plant is also called Indian pipe.

10

Ghost plants are usually white, with black spots. But some are pink.

Ghost plants are not the ghosts of dead plants. And even though they look like mushrooms, they are not. Ghost plants are wildflowers. They grow from seeds.

Ghost plants have an unusual family. Blueberries and cranberries are part of it. Flowers called azaleas are also in the family.

After ghost plants flower, they may develop fruit. The fruit dries out and splits open. The wind blows its tiny seeds away.

The seeds need to land on the perfect spot to grow. They get their energy from just one kind of fungus. This fungus lives in dark, old forests.

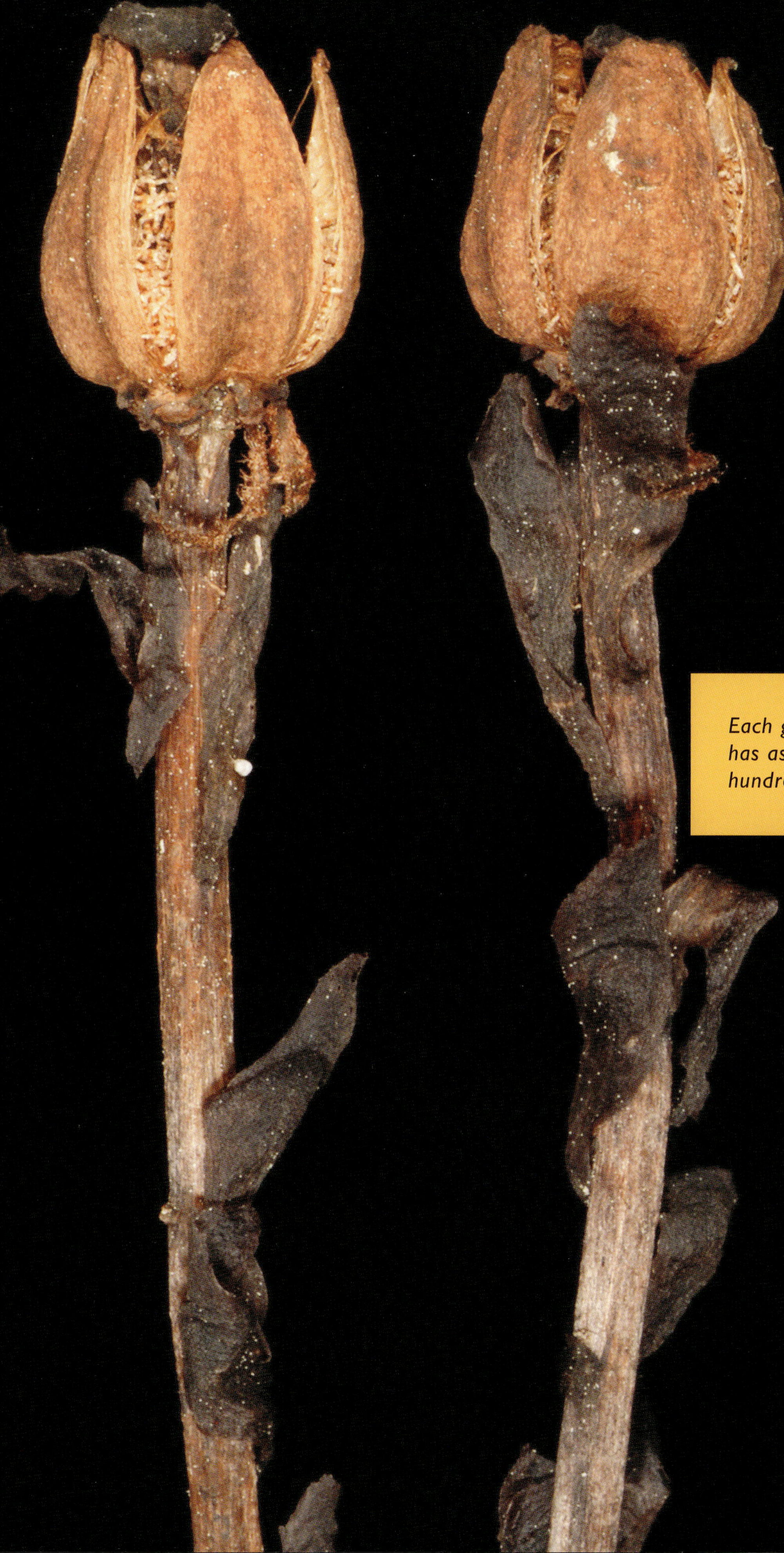

Each ghost plant has as many as a hundred seeds.

Inside the seed is an **embryo**. A hard shell surrounds the embryo. This shell is the seed coat. Water softens the seed coat.

embryo: part of a seed that grows into a plant

Short roots grow from the softened seeds. They reach the fungus and get energy. A stem grows up from the seed.

Ghost plants have short, scaly leaves.

Each plant grows up to 12 inches (30.5 centimeters) tall. It flowers in the summer. There is only one flower per stem. The flowers have no smell.

Bees carry **pollen** from one flower to another. This helps flowers make fruit with seeds. The ghost plant's hard fruit isn't good to eat, though.

pollen: yellow powder made by plants that is used to make other plants

The ghost plant flowers are shaped like bells.

Most plants are green because of **chlorophyll** (KLOR-uh-fil). It helps them get energy from the sun. But ghost plants do not get energy from the sun. They do not need chlorophyll.

People can't grow ghost plants in gardens. Ghost plants are delicate. They turn black if they are touched. They will die without the fungus they need for energy.

chlorophyll: what makes plants green

Ghost plants can grow in the darkest parts of forests.

The ghost plant grows in dark forests. It starts as a seed. A tangle of roots grows from the seed. It finds a fungus and gets energy from it. A white, scaly stem grows. A bell-shaped flower blooms at the end of the stem. After bees bring pollen to the flower, it grows fruit. The fruit dries up. Seeds pop out. The cycle begins again.

LIFE STAGES

Seed
Root
Stem
Flower
Fruit
Seed Pod

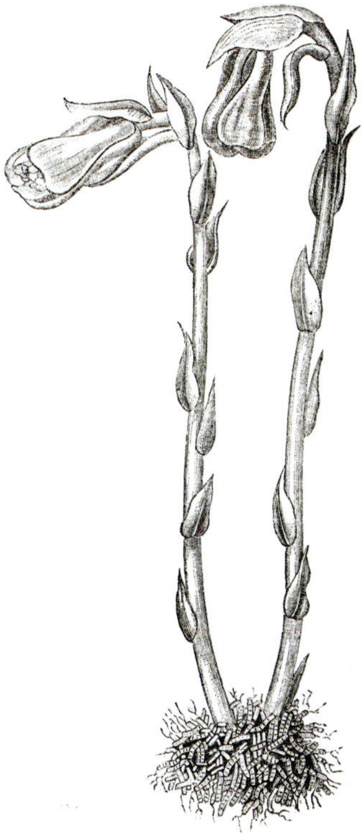

A long time ago, the Cherokee people were happy. They never fought with their neighbors. But then they started arguing. They argued even while they shared a tobacco pipe, a symbol of peace. This angered the Great Spirit. He turned the arguing people into small flowers that looked like pipes. This is why ghost plants are also called Indian pipes.

WEBSITES

Indian Pipe – Friends of Sleeping Bear Dunes
https://friendsofsleepingbear.org/wildflower-database/white/indian-pipe
Learn more about the ghost plant, also known as Indian pipe.

16 weird plants that seem more supernatural than natural
https://www.nhm.ac.uk/discover/weird-plants.html
Read about 16 unusual plants, including the ghost plant.

READ MORE

Einstein, Tamara. *Weird Plants.* Tukwila, WA: KidsWorld Books, 2021.

Lundgren, Julie K. *Creepy but Cool Scary Plants.* New York: Crabtree Seedlings, 2021.

Mattern, Joanne. *Disgusting Plants.* Minnetonka, MN: Kaleidoscope, 2020.

INDEX